MW01320794

WHAT IS GOOD DESIGN?

Other Schiffer Books on Related Subjects:

Alexander Girard Designs for Herman Miller, Leslie Piña, with foreword by Stanley Abercrombie, ISBN 978-0-7643-1579-4

Luxury Design for Living, Steve Huyton, ISBN 978-0-7643-5421-2

Library of Congress Control Number: 2021942715

Cut-paper illustrations by Lisa S. Roberts
Design by Lisa Benn Costigan

Type set in Questrian/Proxima Nova A

ISBN: 978-0-7643-6405-1
Printed in Serbia

Published by Schiffer Publishing, Ltd.
4880 Lower Valley Road
Atglen, PA 19310
Phone: (610) 593-1777; Fax: (610) 593-2002
E-mail: Info@schifferbooks.com
Web: www.schifferbooks.com

WHAT IS GOOD DESIGN?

A SIMPLE QUESTION WITHOUT A SIMPLE ANSWER

BY LISA S. ROBERTS

4880 Lower Valley Road • Atglen, PA 19310

LISA S. ROBERTS IS A LIFELONG STUDENT OF DESIGN

Beginning her career as an architect, she went off-course and became a product designer focusing on objects for the home and gift markets. Many of her designs can be found in stores across the country, particularly museum stores.

Veering off course again, she wrote three books on design and was the host of a TV docuseries that featured her and her design team. She even has a cat, Mr. Waffles, who loves design and was the subject of her third book.

Lisa sits on the board of the Cooper Hewitt Smithsonian Design Museum and the Philadelphia Museum of Art.

She lives, breathes, and dreams of good design.

Images top to bottom: Men's necktie design, Mr. Waffles, Cooper Hewitt Smithsonian Design Museum facade

WHAT IS GOOD DESIGN?

This is a simple question, but there is no simple answer. Everything we use and everything around us, whether it's a bicycle, chair, trash can, or water bottle, has been designed by someone.

There are many things that make something GOOD DESIGN. It can be how it looks or how it works. It can be a new idea that no one has seen before, or improves on the way things were done in the past. However, what makes one thing good design may not be what makes another thing good design—this is why it is not a simple question to answer.

Why should we even care about Good Design? Because it makes life better!

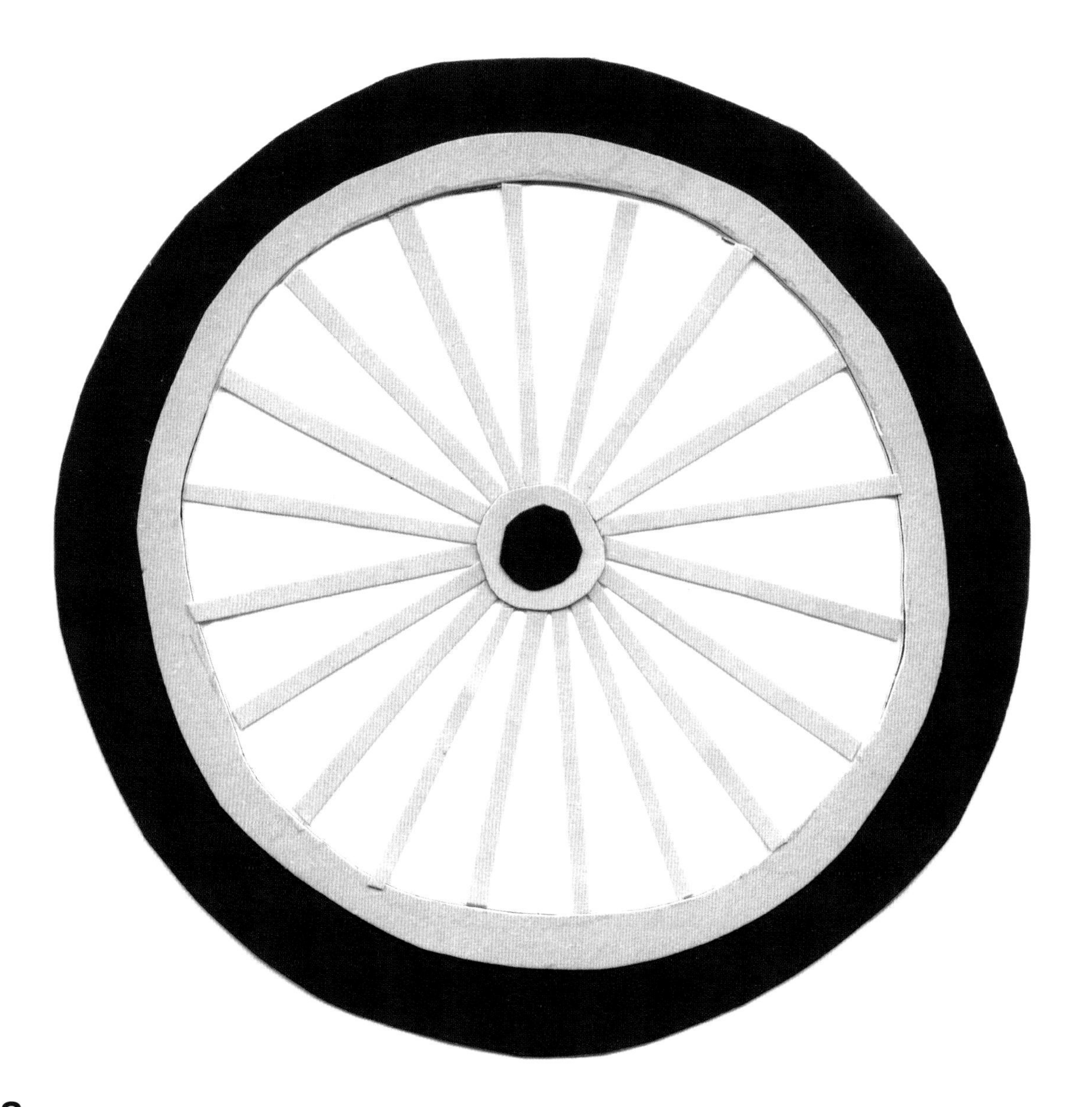

THERE ARE MANY WAYS
TO IDENTIFY GOOD DESIGN.
HERE ARE SEVEN
OF THE MOST IMPORTANT.

1. WORKS WELL
2. LOOKS GOOD
3. OPENS OUR MIND TO NEW IDEAS
4. IMPROVES LIVES
5. GIVES US JOY
6. IS INNOVATIVE
7. STANDS THE TEST OF TIME

WORKS WELL

The objects we use every day have a job to perform. If they do it well, it makes our lives easier, better, and, ultimately, more enjoyable.

For some, how an object functions is the most important aspect of design.

BROMPTON

FOLDING BIKE

Why is the Brompton folding bike better than
the standard commuter bike?

It folds in four places and collapses into a shape that is easy to carry—up stairs, on a bus, or on a train. And because it's easy to store away, there is no need to lock it up outside, where it might get stolen.

OXO GOOD GRIPS

The handles of these kitchen utensils are designed for hands of all shapes and sizes:

—strong and weak
—young and old
—lefties and righties

This is called Universal Design.

MANY MUSEUMS HAVE THIS SWIVEL PEELER IN THEIR COLLECTIONS.

MAY DAY LIGHT

Why did the designer name the light May Day? Mayday is an emergency call for help when you're in trouble at sea. This light is there to help wherever and whenever you need it.

UNIQUE FEATURES

- Portable
- Large, hook-shaped handle for hanging
- Versatile as floor, table, or hanging light
- Large cone provides plenty of light
- Convenient, extra-long 16-foot cord

RADIUS
TOOTHBRUSH

What makes this toothbrush better than most others?

- Large, oval head with three times as many bristles cleans more of your teeth and gums at one time.
- Oversized handle is comfortable to hold and easier to control with a place for the thumb and palm.
- First-ever right-hand and left-hand toothbrushes.

XOX TABLE

Some people think Good Design means something that's made as simple as possible. They even say "less is more." The XOX table definitely fits that description—it has only three parts! Assembled, it becomes a fun and functional table.

1. The pieces slot together—no glue or nails. 2. The table is made of fiberboard (wood fibers). It's inexpensive but so sturdy it can hold a stack of books. 3. It's packed in a flat box and you put it together, so it costs a whole lot less to ship than a ready-made table.

LOOKS GOOD

Not everyone likes the same things, but when it comes to the design of an object, people's opinions are based on similar things: shape and form, colors and textures, and, sometimes, whether it reminds them of something else they love.

EAMES
STORAGE UNIT

Charles and Ray Eames designed this storage unit to be as interesting as a piece of art.

They combined open shelves, closed cabinets, and drawers with different colors, materials, and patterns. They even introduced their new storage unit at an art exhibit rather than in a furniture showroom. The parts can be moved around, so each person gets to choose their own design.

ANTIBODI CHAIR

Two things immediately stand out that make this chair a wonder: the many bright colors and the oversized flowers. The material also adds to its appeal. It's made of felted wool, which is soft and inviting. To relax on this bed of blossoms is a truly enchanting experience.

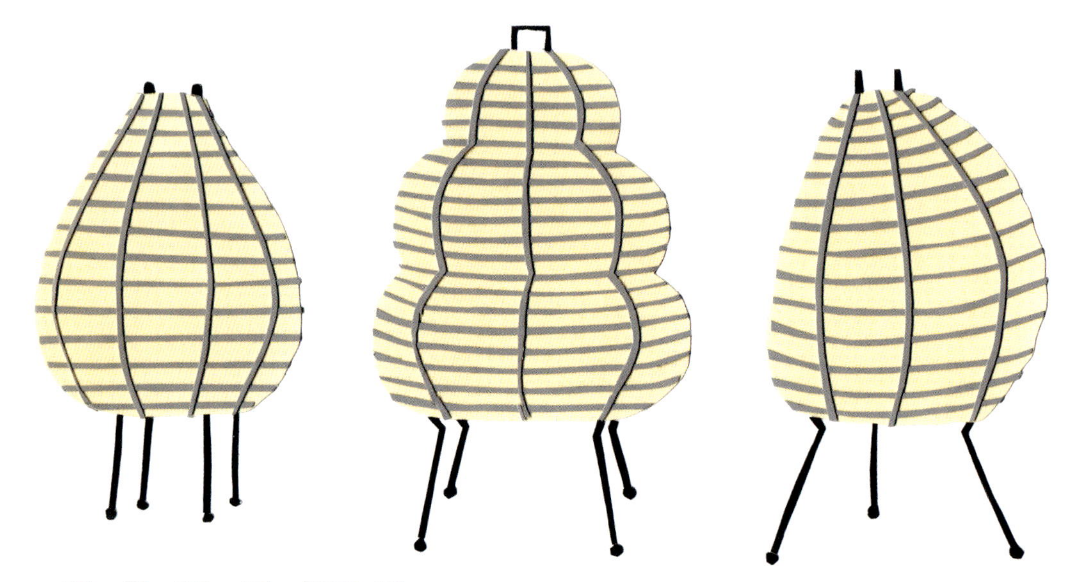

AKARI
LIGHT SCULPTURES

Isamu Noguchi was an artist and sculptor who later became a designer. His Akari table lamps look more like sculptures than traditional table lamps. They are made of translucent Japanese paper wrapped around organically shaped frames. Clearly, he intended them not just to provide light but to be works of art for the table.

MARSHMALLOW SOFA

What is it about this sofa that is so fun and appealing?

1. **The colors.** It has many bright colors—most sofas are all one color.

2. **The shape.** This sofa is made up of round circles, while most sofas have a rectangular shape.

3. **The name.** Who wouldn't love a sofa called Marshmallow? Yum!

TOWER
SALT & PEPPER GRINDERS

Everyone uses a salt and pepper shaker—but how many are actually interesting? These are! They look like two colorful guests who were invited to dinner. The magic is how you use them. They are not really shakers but grinders. With a twist of their heads, salt comes out of the little one and pepper comes out of the tall one. When the meal is finished, you don't have to put them back in the cabinet because they just look good hanging out on the table.

OPENS OUR MIND TO NEW IDEAS

Designers can change our expectations about the way something should look. They can create a totally different design that we might not even recognize. This opens up new possibilities and improvements on the way things were done before!

EXCALIBUR
MERDOLINO
TOQ

TOILET BRUSHES

have always been just a brush on a stick.

Not this group!

These brushes bear no resemblance to their predecessors. They amuse and even entertain us while we scrub the bowl.

LOOKS LIKE A SWORD

LOOKS LIKE A POTTED PLANT

LOOKS LIKE A ONE-EYED MONSTER

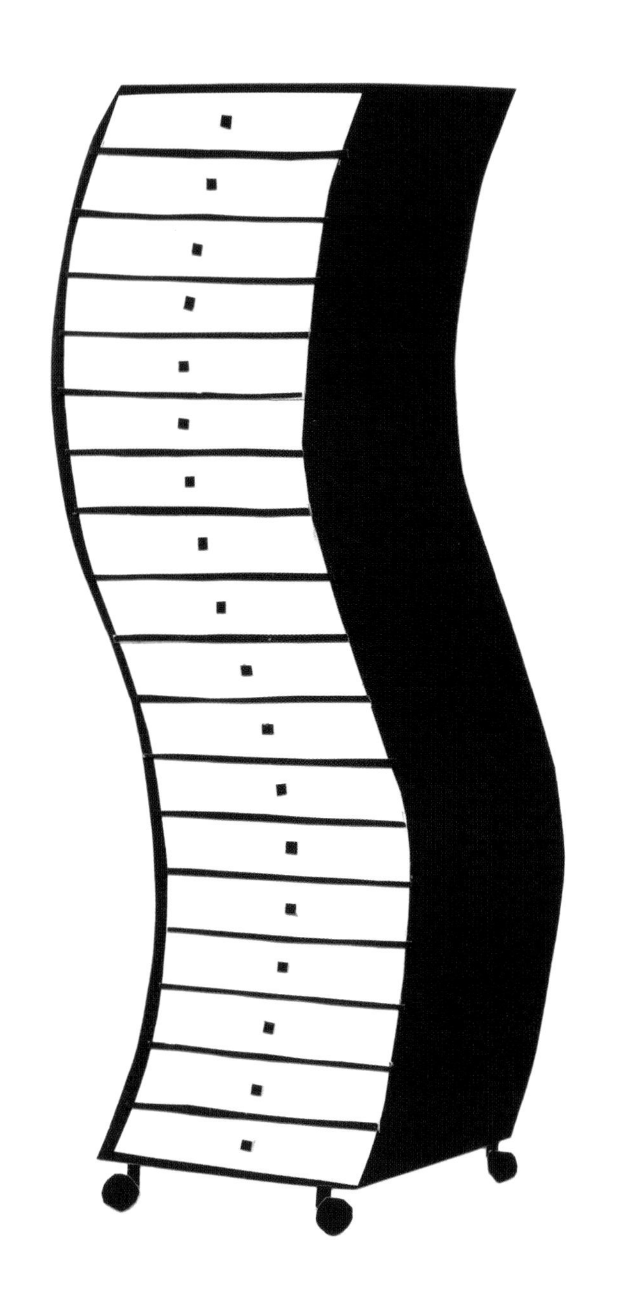

SIDE 1

CHEST OF DRAWERS

Most chests of drawers are boxes
They may be tall or wide, but rarely wavy
Japanese designer Shiro Kuramata broke away
from the traditional shape and designed
a chest of drawers that undulates
as though it were alive
The drawers pull out smoothly and hold
everything they should, so the design is functional
But it's the unexpected shape that makes us
pay attention to it and wonder, How does it do that?

MAX HEATER

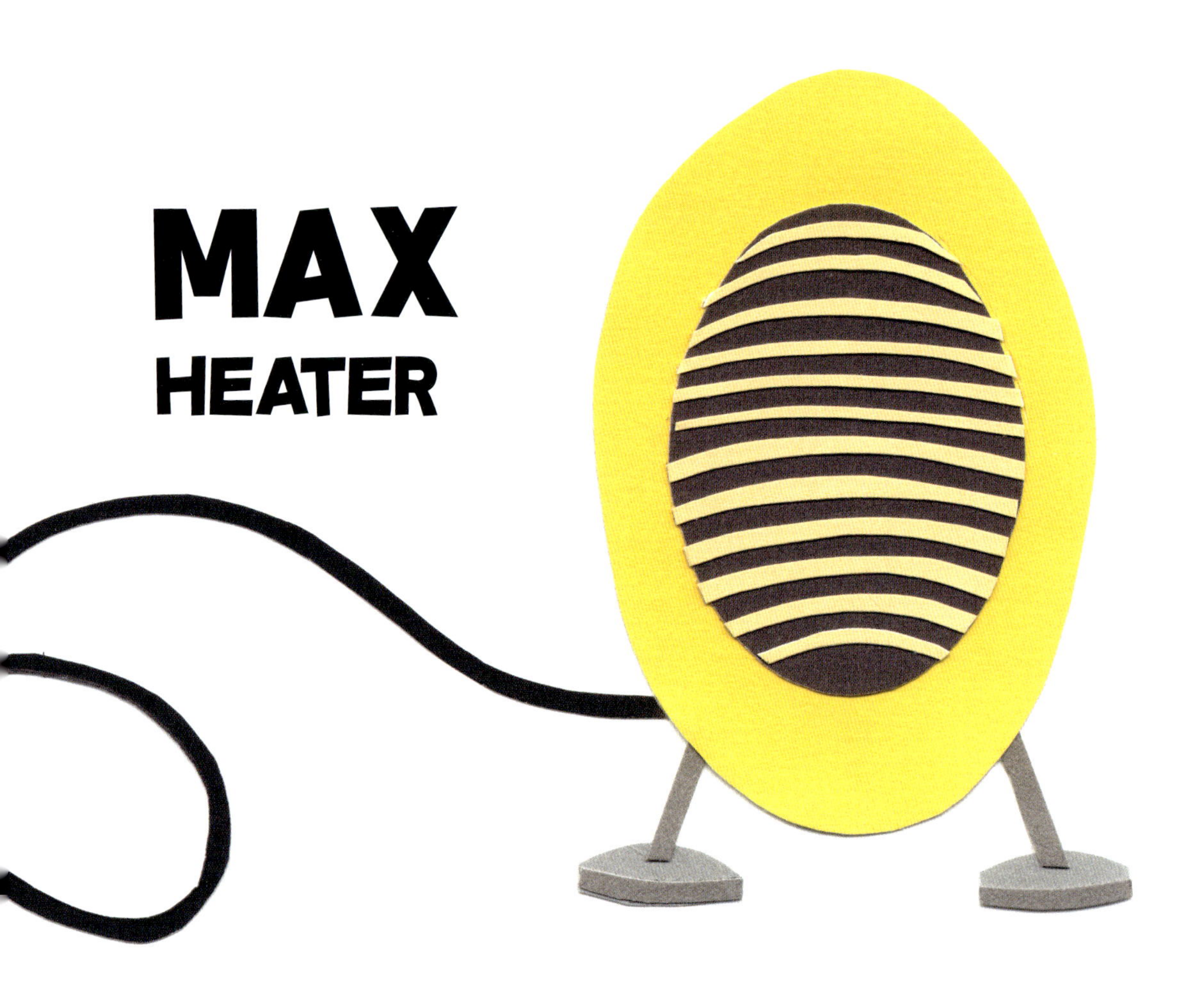

Most portable heaters are basic black, white, or gray boxes. Max Heater is surprisingly different. He works just like a regular heater but looks like he may have landed from another planet. His bright-yellow body, three stable feet, and gentle purring noise almost make you think he's alive.

Like his brother, Fred looks nothing like other humidifiers. His round body, legs, and snout resemble a pet. Like the family dog, he provides comfort: warm steam shoots from the snout and three small lights let you know he's there and keep you from stepping on him when getting out of bed.

FRED

HUMIDIFIER

JUICY SALIF

LEMON JUICER

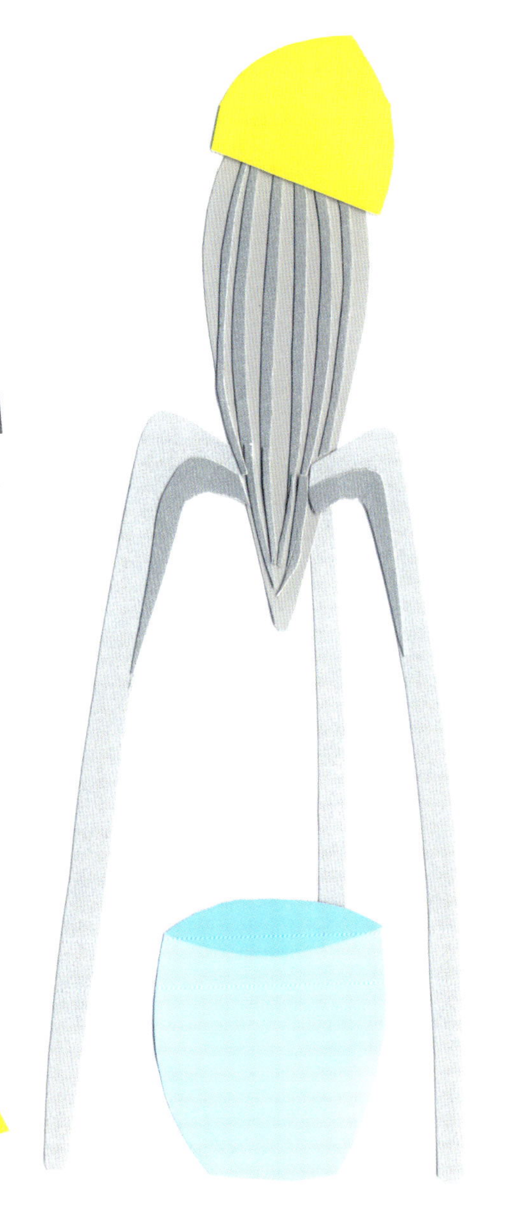

At first glance, no one would guess this is a lemon juicer. The designer Philippe Starck loved surprising people by totally changing their expectations about the way something should look. It is sculptural in form—he wanted you to treat it like a work of art and keep it on the counter instead of in a drawer. People loved it and did just that!

WHERE DID HE GET THE IDEA?

One day he was eating fried calamari (similar to octopus) in a restaurant. It came with a piece of lemon to squeeze on top. He took out a pencil and started to sketch a design on his place mat. He thought about what he was eating, put the two together, and came up with a lemon juicer in the shape of an octopus!

WIGGLE CHAIR

* Made of 53 sheets of cardboard glued together.

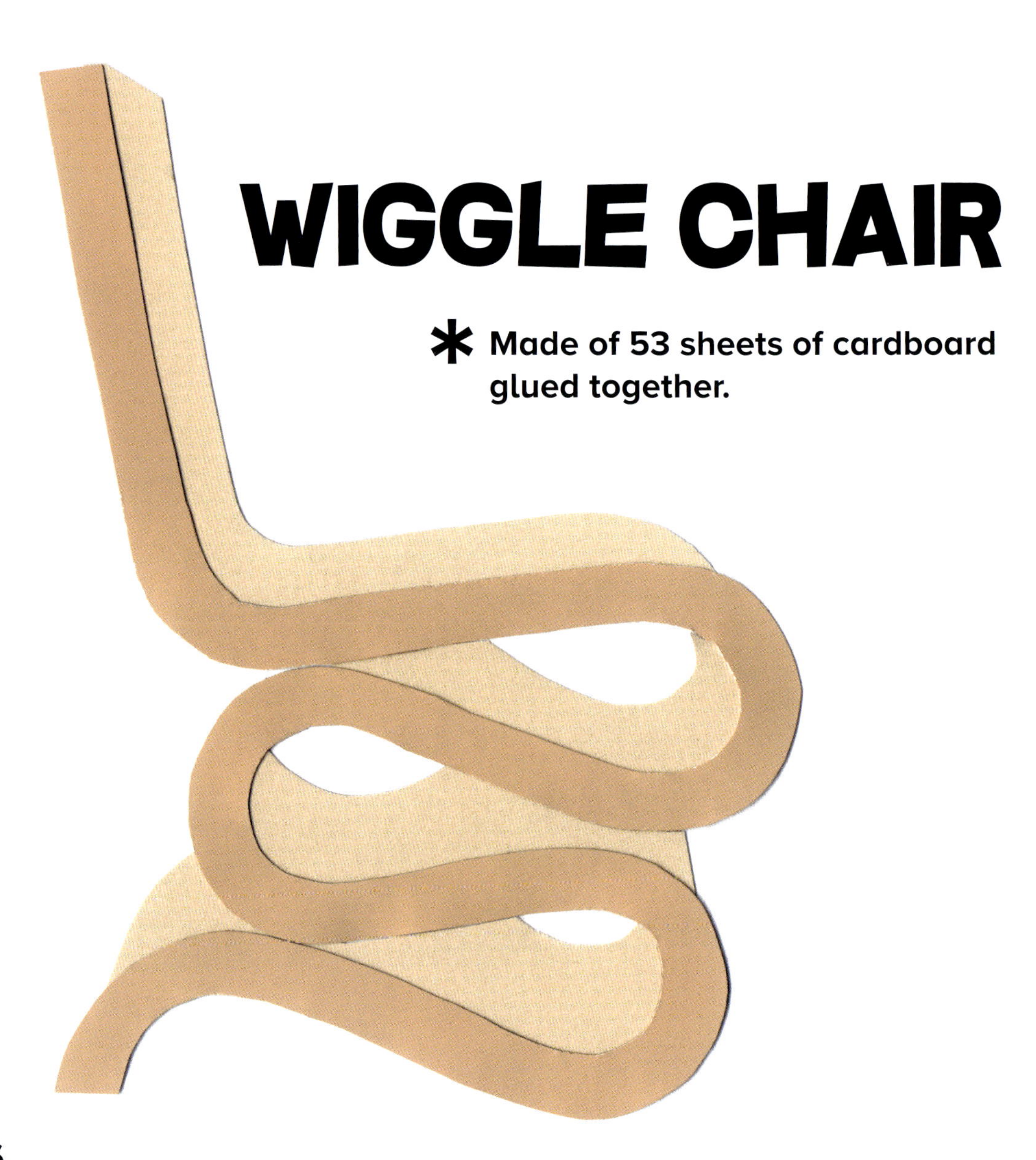

Corrugated cardboard is usually used for boxes, not furniture. It's just not sturdy enough on the flat side to hold a lot of weight. However, if you turn it on its vertical side and stack sheets of it together, it becomes structural. That means it can hold a lot more weight and can be used for a lot more things.

[Close-up view of layers]

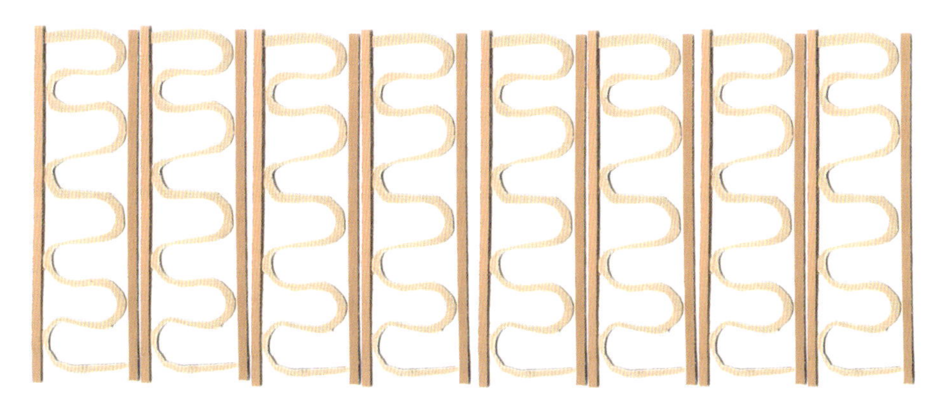

Good Design does not have to be made from expensive materials. With creativity, the most ordinary material can be transformed into amazing designs.

IMPROVES OUR LIVES

Good design can make our lives easier, safer, healthier, faster, simpler, or, in general, just plain better.

LITTLE SUN

This is a personal solar light that can be used as a flashlight, reading light, or night light. The light is on the front and, when it is charged, can give up to 50 hours of light. The solar panel is on the back and can charge in sunny or cloudy weather.

The creators of Little Sun care about people who don't have electricity. For every one sold, another one goes to a community in Africa at a much-lower price.

VICTORINOX

SWISS ARMY KNIFE

The original knife was created 120 years ago to be a compact knife with a lot of different functions. Over the years, those functions have changed and expanded. One model has 33 tools that include blades, a bottle opener, screwdrivers, saws, a corkscrew, scissors, pliers, a magnifying glass, a fish scaler, a nail file, a ballpoint pen, a ruler, even a toothpick! The company makes 100,000 knives a day, and they are so useful, they have saved lives on land and water and in the air.

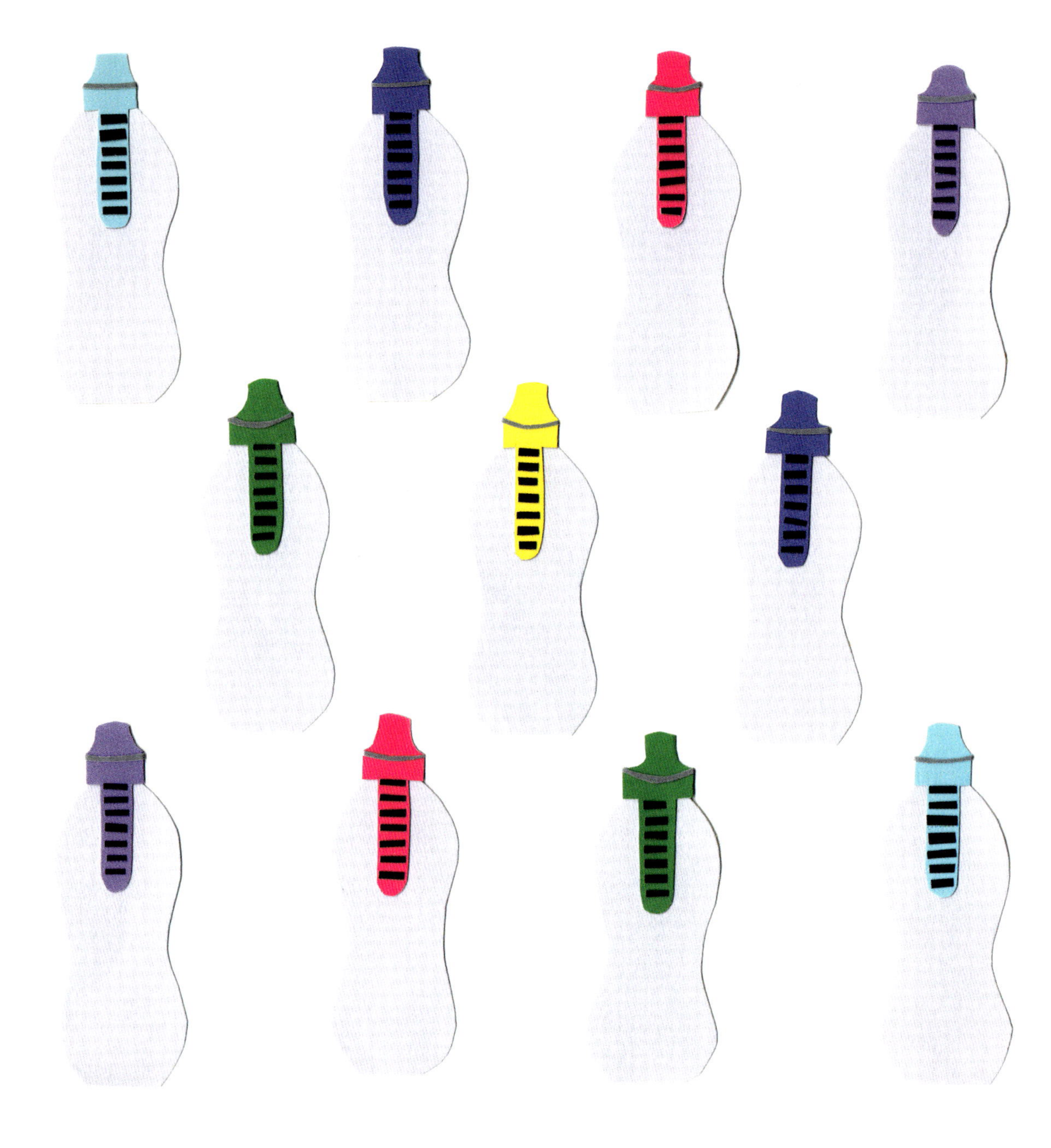

BOBBLE WATER BOTTLE

This is the first water bottle to have a built-in water filter. It is made out of charcoal and can filter out chlorine, hard metals, and other things that make water taste and smell bad.

One filter can be used up to 300 times, saving 300 bottles from being thrown away.

The bottle also has other good design features. Its curvy shape makes it easy to grip, and it comes in a lot of colors so it's easier to identify which is yours.

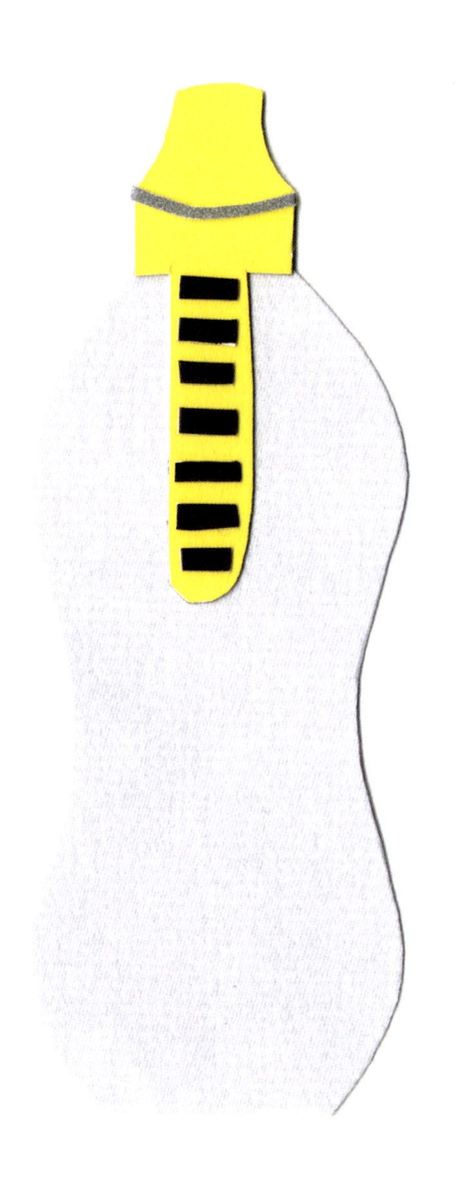

LUCKY IRON FISH

Iron is important for our body because it helps make our blood. Without it, we can get anemia, which makes us tired and weak. Many people do not have enough iron because they don't get the right foods, and iron pills can be expensive. There is another solution—adding a piece of iron to a cooking pot.

The Lucky Iron Fish is just that, a piece of iron metal in the shape of a fish that goes inside a pot of water. When it boils, it releases the iron in the food or water.

Why is it in the shape of a fish? The inventor was trying to help people in Cambodia who needed iron in their diets. His first design was a square piece of iron, but nobody wanted to put that in their pot. Then, he designed it in the shape of a fish because it was their symbol of good luck. Now, people love adding them to their cooking pots. It was a simple solution that improved the health of many people.

LUMOS MATRIX HELMET

The Lumos Matrix helmet is not like other helmets. It has lights built into both the front and back. The front light is a headlight that helps you see bumps in the road and gets the attention of cars and pedestrians.

The back of the helmet has a "matrix" or grid of lights that you can control. When they are all lit, it means you are stopping. It can display arrows to show you are turning left or right. It can even light up to say words like "Hello" or "Stop."

This helmet protects bikers if they fall, but it also makes them more visible day and night. It might even save their life!

5

GIVES US JOY!

Joy is not something you can touch, but something you might feel when you see or use an object. It can put a smile on your face or warmth in your heart.

Is this a bath toy or maybe a plaything for the beach?

NEITHER!

SHIP SHAPE

BUTTER DISH

This fun little boat can be found on the dining table.
Pull on the "smoke stack," and surprise!
A butter dish in disguise!

WHISTLING BIRD
TEA KETTLE

Architects mostly design buildings; however, architect Michael Graves also designed lots of things for the home. His tea kettle was his most famous design. He realized most tea kettles are quite boring, so he created this one to make us smile.

It has a little red bird whistle that "sings" when the water boils; a pretty blue, curvy handle that fits comfortably in the hand; and a black ball that makes it easy to lift off the top.

His tea kettle is probably the most successful one ever made. Over TWO MILLION have been sold, totaling $350,000,000!!!

CARLTON

ROOM DIVIDER

Practically speaking, Carlton is a bookshelf with drawers.
But practicality is not its purpose.
Bright colors, crazy angles,
and a little man standing on top create
"furniture drama."
It doesn't even need any books or knickknacks—
it makes you happy all by itself.

What a shock to open up an umbrella on a dreary, rainy day and suddenly see blue skies and fluffy white clouds. You were not expecting this, because when it's folded it looks just like any other ordinary black umbrella. However, the surprise of opening it and seeing something impossible—blue skies on a rainy day—gives us a spark of joy.

If rainy days get you down, just open this umbrella
and pretend you're somewhere else.

SKY UMBRELLA

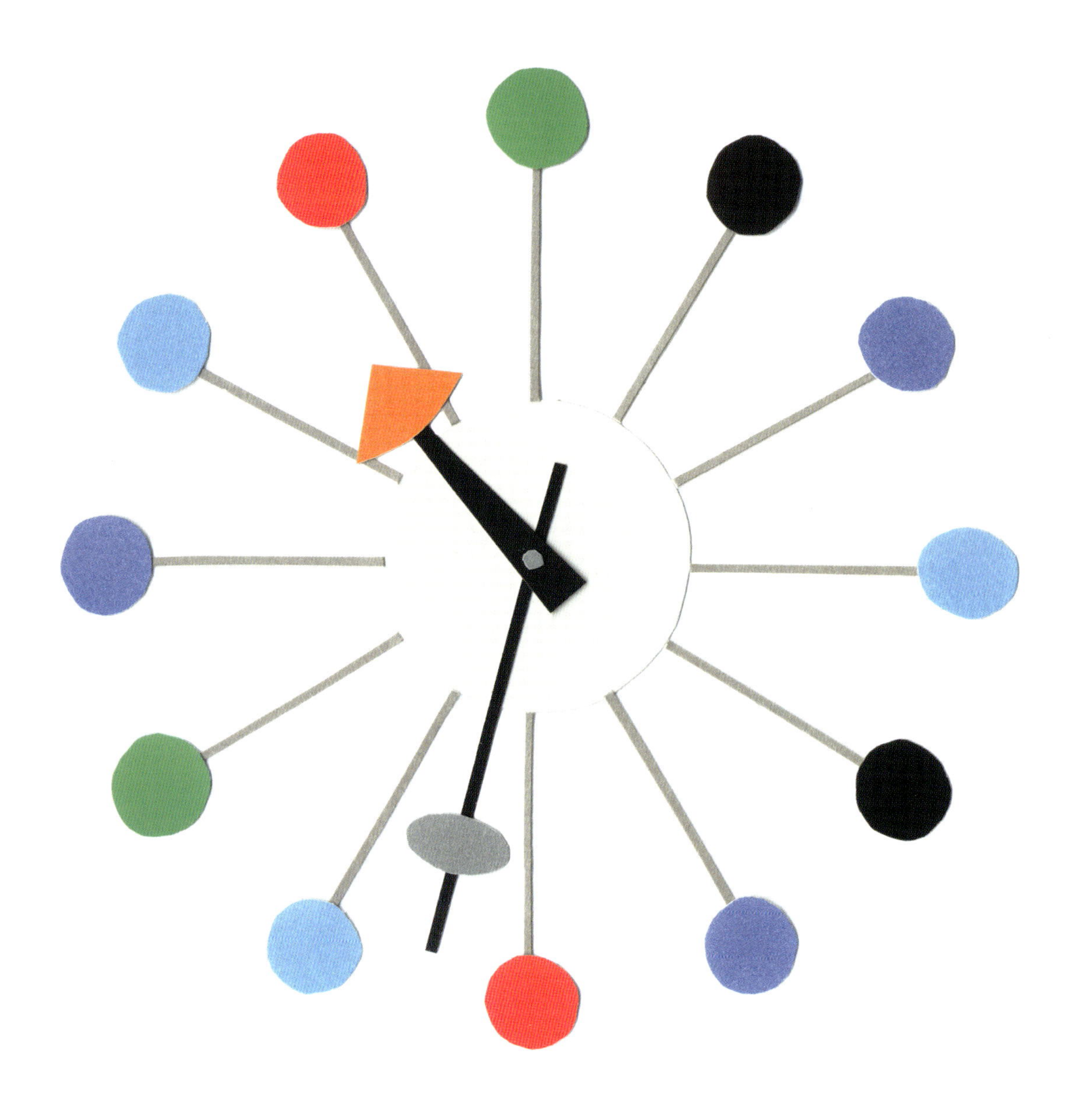

BALL CLOCK

The Ball Clock looks more like wall art than a wall clock. There are no numbers, just colorful spheres and two hands. Most people can tell time without the numbers just by looking at the position of the hands, so it didn't matter that they were left out.

The clock was designed in the 1940s, when there was a lot of interest in outer space. Designers used stars, asterisks (*), and atoms (balls going around a circle) in their designs. The Ball Clock combined some of these shapes in its unique design. Simple, colorful, and fun, this is a happy clock, no matter what time it is.

IS INNOVATIVE

Innovation in design successfully solves a problem in a new way or transforms an older design to make it much better.

BOOKWORM

BOOKSHELF

Ron Arad's flexible bookshelf is a strip of plastic that can be shaped into a spiral, a worm, an S shape, or just about anything sinuous. The more bends and curves in the configuration, the more weight the shelf can bear.

Unlike most other bookshelves, the one shape it cannot take is a straight line.

DYSON VACUUM CLEANER

Vacuum cleaners typically have bags in them that collect the dust and dirt. But they can get clogged very easily. Engineer James Dyson solved that problem when he invented the first BAGLESS vacuum cleaner.

But he didn't stop there—he made a lot of other improvements to the standard vacuum cleaner. First, he designed an electric motor that greatly increased the vacuum's suction power. Second, he added a roller ball that allows the brush head to twist and turn in all directions to make vacuuming easier. He also knew that most people don't like to vacuum, and that it might be more appealing if the machines came in fun colors. So, he made them in yellow, pink, purple, orange, red, and more.

Arm lifts
out and
extends
to vacuum
hard-to-reach
corners
See-through
canister
unscrews
to empty
dirt and dust
Rollerball
can turn
360 degrees
Rotates to
clean floor
or carpet

PLUMEN
LIGHTBULB

Plumen is the world's first designer lightbulb. Its shape is unlike any other.

The classic bulb is a simple form that looks the same from every angle. Plumen looks different from every angle, even though it is the same bulb.

It screws into a socket like a traditional incandescent bulb but uses today's energy-saving LED technology, and the bulb doesn't get hot.

Plumen is also attractive—no need for a lamp shade to cover it up.

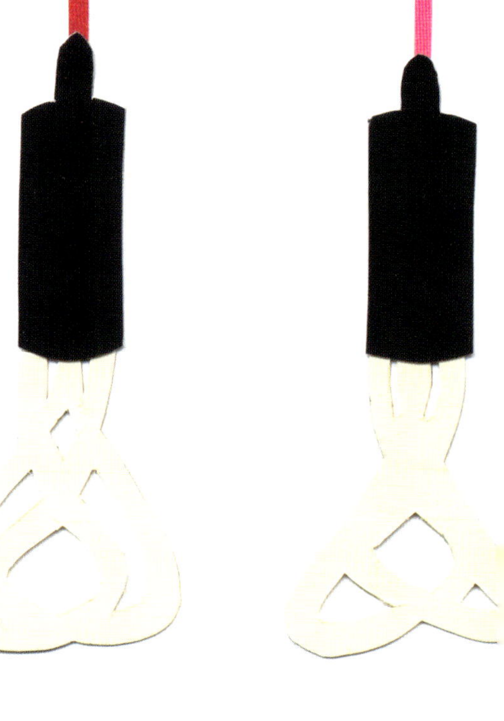

The Plumen lightbulb shown from different angles

iPHONE

In 2007, the first-generation iPhone was invented, and it was revolutionary. The design was simple on the outside—but complex on the inside. It was the first cell phone that had a built-in MP3 player (which stored and played music) and computer all in one. There was no keyboard or pointer and no buttons—just a single "home" button and touch screen.

Over time, iPhones got better and better. Today, you can FaceTime with friends, text, take photos and make videos, share music, play games, ask Siri questions, get directions, and translate languages. And that's just a few of its capabilities! What can't you do with an iPhone? *

*** Wash the dishes, walk the dog, or make your dinner.**

SPORK

Just like the name sounds, a spork is a combination spoon and fork. It's been around for almost 150 years! Its creator was not a designer but a medical doctor who dabbled in inventing things. Sometimes the best designs seem the most obvious, and indeed this simple object makes perfect sense. There is one less utensil on the table and one less to clean and store. Today, the design has been changed a little to make them lighter and more portable—most sporks are made of plastic or wood instead of metal. Sporks are handy to use at fast-food restaurants and schools, on picnics and airplanes, in backpacks—anywhere you're on the go.

7

STANDS THE TEST OF TIME

Trends come and go, but some objects never go out of style because they work so well or look so good, and people continue to love them. They've become a classic.

FISKARS SCISSORS

When they were created in 1967, Fiskars were the first scissors to have plastic handles. Others were metal, making them much heavier. The shape of the handle is ergonomic. It fits the hand like a glove, which makes cutting easy. The design has changed slightly over time—but it still has the famous orange handle. **Over 1 billion have been sold!**

ANGLEPOISE LAMP

George Carwardine was a car designer before he created this lamp. He worked on suspension systems that used springs and shock absorbers to support cars on the road. He saw that springs could also be used for other things, and designed this lamp with them.

The lamp is made of different sections that can move up and down into many lighting positions. Three springs at the bottom create tension in the arms of the lamp. This keeps it in perfect balance so that it holds any position.

Anglepoise was created in 1935, and the design hasn't changed since. When it's perfect the first time, it doesn't need improvement!

PANTON STACKING SIDE CHAIR

SAARINEN TULIP CHAIR

MIDCENTURY FURNITURE was created between the 1930s and the 1960s. Why is it still popular? Before that time, furniture tended to be big, heavy, and decorative. During the 1930s, there was a new interest in making furniture lighter, more streamlined, and easier to produce.

Also, new materials such as molded plastics and plywood made furniture comfortable and affordable. Today, people still like the clean lines and simplicity of midcentury modern furniture, which is why it has never gone out of style.

WEBER KETTLE GRILL

This is perhaps the most popular grill ever made, first designed 70 years ago. Before it was invented, most grills were just flat iron bars. The food had to be turned back and forth so it cooked all the way through. The round grill solved a lot of problems. The lid keeps the heat in and spreads it out evenly so the food cooks on all sides. Small vents at the top and bottom allow the chef to control the heat. This grill is easy to use, great for cooking, and inexpensive—all the ingredients of a design that stands the test of time.

Just how popular is this grill?
The official emoji® for a grill is the Weber Kettle Grill.

Legos are so popular that there are Lego amusement parks, Lego movies, Lego games, and Lego books. The Lego world keeps getting bigger and bigger!

LEGOS

Legos might seem like an odd choice to have in a book about Good Design. But it may be one of the best examples. The concept is simple. One Lego brick connects to another identical brick with pegs on one side and holes on the other. Any number of things can be built using this singular element, limited only by your imagination. The company later added new shapes and sizes so you could make many more things, such as trains, buildings, dollhouses, action figures, dinosaurs, monsters, space ships—you name it!

The first Lego bricks were created in 1932 and made out of wood. They were heavy and expensive. Then the company switched to plastic, which made the blocks lighter, easier to use, and a lot less expensive—another example of good design.

WHAT IS

GOOD DESIGN?

Now that you've looked through these pages, you can see that there are many ways to understand good design. A tea kettle that looks good might also be a delight to use. A toilet brush that resembles a one-eyed monster might expand your ideas about how a toilet brush should look, and also make you laugh while you're doing an unpleasant job. A pocketknife that improves your life, and might someday save it, has stood the test of time because of the ingenious way it's put together. And a vacuum cleaner that is inventive might also have striking good looks, work well, and improve your life, all at the same time.

Answering the question
"What is good design?"
is not always simple, but
figuring it out means making
better choices about the
things we have around us.

What good designs are in your life?

WHO
WHAT
WHEN

Anglepoise Lamp
Designer: George Carwardine
Manufacturer: Herbert Terry & Sons
Date: 1932
Materials: various metals

Antibodi Chair
Designer: Patricia Urquiola
Manufacturer: Moroso
Date: 2006
Materials: wool felt, stainless steel

Akari Light Sculptures
Designer: Isamu Noguchi
Manufacturer: Ozeki & Co.
Date: 1951
Materials: washi paper, bamboo, steel

Ball Clock
Designer: Irving Harper / George Nelson
Manufacturer: Vitra
Date: 1947 (design) 1949 (production)
Materials: beech, metal, and acrylic

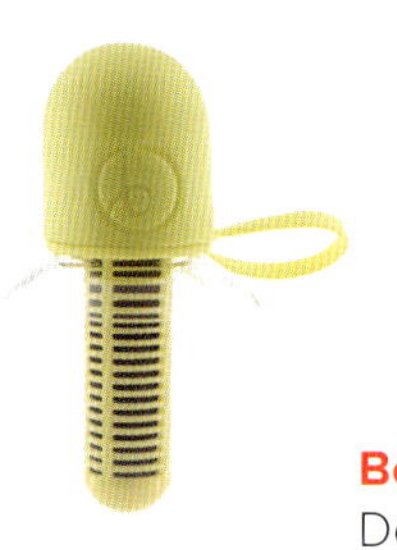

Bookworm Bookshelf
Designer: Ron Arad
Manufacturer: Kartell
Date: 1994
Materials: PVC plastic

Bobble Water Bottles
Designer: Karim Rashid
Manufacturer: Move Collective, LLC
Date: 2010
Materials: recycled plastic, carbon

Brompton Folding Bicycle
Designer: Andrew Ritchie
Manufacturer: Brompton Bicycle
Date: 1979
Materials: steel, titanium, rubber, others

Carlton Room Divider
Designer: Ettore Sottsass
Manufacturer: Memphis Milano
Date: 1981
Materials: wood, plastic laminate

Dyson Vacuum Cleaner
Designer: James Dyson
Manufacturer: Dyson Ltd.
Date: 1991–present
Materials: variable

Eames Lounge Chair Wood (LCW)
Designers: Charles and Ray Eames
Manufacturer: Herman Miller
Date: 1940 (original design)
Materials: molded plywood

Eames Molded Plastic Armchair
Designers: Charles and Ray Eames
Manufacturer: Herman Miller
Date: 1948–50
Materials: fiberglass, metal base

Eames Storage Unit (ESU)
Designers: Charles and Ray Eames
Manufacturer: Herman Miller
Date: 1950
Materials: plywood, lacquered Masonite, steel

Excalibur
Designer: Philippe Starck
Manufacturer: Heller
Date: 1993
Materials: polypropylene, nylon

Fred Humidifier
Designer: Matti Walker
Manufacturer: Stadler Form
Date: 2000
Materials: plastic, chrome-plated zinc

Fiskars Scissors
Designer: Olof Bäckström
Manufacturer: Fiskars Oy Ab
Date: 1967
Materials: plastic, stainless steel

iPhone
Designer: Apple / Jonathan Ive
Manufacturer: Apple
Date: 2007
Materials: glass, metal, plastic

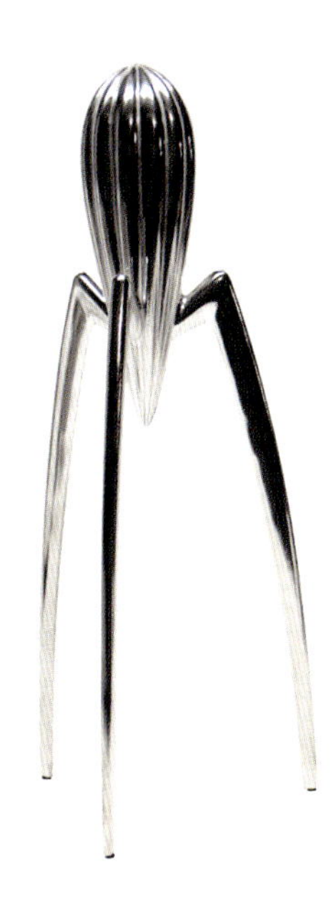

Juicy Salif
Designer: Philippe Starck
Manufacturer: Alessi
Date: 1990
Materials: aluminum

Legos
Designer: Ole Kirk Christiansen
Manufacturer: Lego
Date: 1958
Materials: ABS plastic

Little Sun
Designer: Olafur Elliasson
Manufacturer: Little Sun
Date: 2012
Materials: polycarbonate plastic, solar panel

Lucky Iron Fish
Designer: Christopher Charles
Manufacturer: Lucky Iron Fish
Date: 2008 (company formed in 2012)
Materials: cast iron

Lumos Matrix Helmet
Designer: Bilal Raja
Manufacturer: Lumos
Date: 2019
Materials: EPS and ABS

Marshmallow Sofa
Designer: Irving Harper
Manufacturer: Herman Miller
Date: 1956
Materials: steel, foam, upholstery

May Day Portable Lamp
Designer: Konstantin Grcic
Manufacturer: FLOS
Date: 1999
Materials: polypropylene plastic

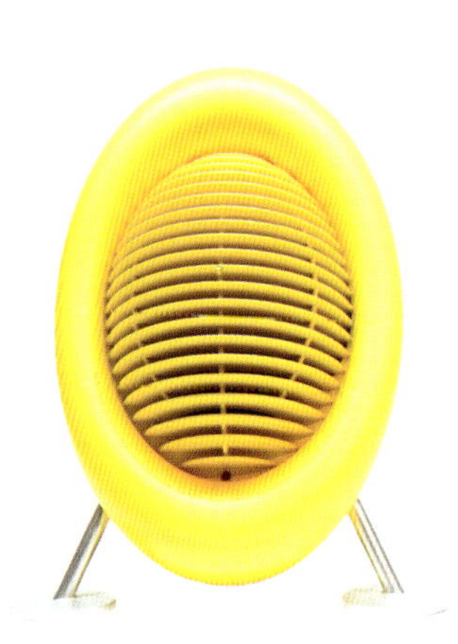

Max Heater
Designer: Matti Walker
Manufacturer: Stadler Form
Date: 2001
Materials: plastic, chrome-plated zinc

Merdolino
Designer: Stefano Giovannoni
Manufacturer: Alessi
Date: 1993
Materials: Thermoplastic resin

OXO Good Grips Kitchen Utensils
Designer: Smart Design
Manufacturer: OXO
Date: 1989–present
Materials: rubber, nylon, silicone, stainless steel

Plumen Lightbulbs
Designer: Nicolas Roope
Manufacturer: Hulger
Date: 2010
Materials: glass, ABS plastics, wires

Radius Toothbrush
Designer: James O'Halloran
Manufacturer: Radius
Date: 1983
Materials: cellulose, nylon

Side 1 Chest of Drawers
Designer: Shiro Kuramata
Manufacturer: Cappellini
Date: 1970
Materials: oak, aluminum, wood

Ship Shape Butter Dish
Designer: Stefano Giovannoni
Manufacturer: Alessi
Date: 1998
Materials: plastic, stainless steel

Sky Umbrella
Designers: Tibor Kalman and Emanuela Frattini Magnusson
Manufacturer: The Museum of Modern Art
Date: 1992
Materials: nylon, aluminum, wood

Spork
Designer: Dr. Samuel W. Francis
Manufacturer: variable
Date: 1874
Materials: variable

Stacking Side Chair
Designer: Verner Panton
Manufacturer: Vitra
Date: 1959–60
Materials: Fiber-reinforced plastic

Tower Salt & Pepper Grinders
Designer: Tom Dixon
Manufacturer: Tom Dixon Studio
Date: 2013
Materials: beech wood, ceramic

TOQ
Designers: Robin Platt & Cairn Young
Manufacturer: Koziol
Date: 2001
Materials: Thermoplastic

Tulip Chair
Designer: Eero Saarinen
Manufacturer: Knoll
Date: 1956
Materials: fiberglass, aluminum, fabric

Victorinox Swiss Army Knife
Designer: Karl Elsener
Manufacturer: Victorinox
Date: 1891–present
Materials: stainless steel

Weber Kettle Grill
Designer: George Stephen
Manufacturer: Weber
Date: 1952
Materials: steel with porcelain enamel

Wiggle Chair
Designer: Frank Gehry
Manufacturer: Vitra
Date: 1972
Materials: corrugated cardboard, fiberboard

Whistling Bird Tea Kettle
Designer: Michael Graves
Manufacturer: Alessi
Date: 1985
Materials: stainless steel, polyamide

XOX Coffee Table
Designer: Josh Owen
Manufacturer: Bozart
Date: 2002
Materials: lacquered MDF

design matters!